Sustainable Packaging Symposium (SPS 2020)

Online
29 June – 1 July 2020

ISBN: 978-1-7138-1590-7

Printed from e-media with permission by:

Curran Associates, Inc.
57 Morehouse Lane
Red Hook, NY 12571

CURRAN ASSOCIATES INC. proceedings .com

Some format issues inherent in the e-media version may also appear in this print version.

AIChE
120 Wall Street, FL 23
New York, NY 10005-4020

Phone: (800) 242-4363
Fax: (203) 775-5177

www.aiche.org

Additional copies of this publication are available from:

Curran Associates, Inc.
57 Morehouse Lane
Red Hook, NY 12571 USA
Phone: 845-758-0400
Fax: 845-758-2633
Email: curran@proceedings.com
Web: www.proceedings.com

TABLE OF CONTENTS

ABSTRACTS

POSTERS

Anaerobic Biodegradation of a Poly (lactic acid) Blend Containing a Poly (lactic acid)-Co-Poly (glycolic acid) Copolymer

Conference: Sustainable Packaging Symposium (SPS)

Year: 2020

Proceeding: Sustainable Packaging Symposium 2020

Group: Abstract Submissions

Session: Panel - Novel Products and Technologies

Time: Thursday, July 2, 2020 - 4:10pm-4:35pm

Authors:

Lewis, C., RIT
Samadi, K., Rochester Institute of Technology
Francisco, M., Rochester Institute of Technology
Trabold, T. A., RIT
Diaz, C. A., RIT
Dell, E., Rochester Institute of Technology
Hegde, S., RIT

Plastics usage is ever increasing due to its many advantages over competing materials including low cost, low density and ability to be converted into useful objects. However, the inability of traditional plastics to degrade results in the accumulation of large amounts of waste. Though recycling remains an attractive option for many applications, mixed waste streams containing plastics co-mingled with other waste types (e.g. food waste) require separation and cleaning steps that renders this option less attractive. Biodegradable plastic materials may offer a solution to this problem providing alternative end-of-life scenarios (e.g. compositing and anaerobic degradation). Unfortunately, many of these materials still exhibit slow degradation kinetics thus limiting this option.

The goal of this work is to examine the anaerobic biodegradation of polylactic acid (PLA) blended with a highly, degradable poly(lactic acid-co-glycolic acid) (PLGA) copolymer. In this experiment both blend composition and PLGA molecular weight were varied. PLA and PLGA were found to be miscible over the molecular weight and blend ratio ranges investigated. The rate of methane production increased with PLGA content in the blend and this was particularly pronounced at early stages of the anaerobic degradation process. Moreover, the rate of methane production increased as PLGA molecular weight decreased. It is envisioned that this strategy could be applied to improve the efficiency of the anaerobic degradation of waste stream containing single-use plastics co-mingled with food waste.

Can Biodegradable Plastics Alleviate Single-Use Plastic Waste? Challenges and Opportunities in the Sustainable Packaging Sector

Conference: Sustainable Packaging Symposium (SPS)

Year: 2020

Proceeding: Sustainable Packaging Symposium 2020

Group: Abstract Submissions

Session: Technical Session - Novel Products and Technology

Time: Thursday, July 2, 2020 - 3:00pm-3:25pm

Authors:

Misra, M., University of Guelph

Plastics are used predominantly in the packaging sector because of their lightweight nature, convenience, and their protective capabilities of packaged products that have a required shelf-life. The world produces around 450 million tons/year of plastic, which is expected to surpass one billion tons/year by 2050. We create 300 metric tons/year of plastic waste that is nearly equivalent to the total human population as per a report by the UN. We need innovations and new approaches to alleviate the plastic waste concern. In addition, fossil fuel plastics represent 99% of plastics sold today and transitioning to green plastics would support the reduction of green house gas (GHG) emissions. The Recycling, reduce, reuse and resource recovery are some of the strategies for waste management. This presentation will highlight the pros and cons of bioplastics in reducing the plastic waste, especially the challenges and opportunities of biodegradable plastics in the packaging sectors. The main barrier to bioplastic adoption today is the price, compared to fossil fuel-based polymers. A cost-competitive and superior performance packaging can be manufactured through innovative compatibilized biodegradable blends and their bio-nano-composites.

Acknowledgements: i) Natural Sciences and Engineering Research Council (NSERC), Canada Discovery Grants (Project # 400320); ii) the Ontario Ministry of Agriculture, Food and Rural Affairs (OMAFRA) â€“ University of Guelph, the Bioeconomy Industrial Uses Research Program Theme (030177 and 030486); iii) OMAFRA-University of Guelph Gryphonâ€™s Leading to the Accelerated Adoption of Innovative Research (LAAIR) Program (Project # 298635); iv) the Ontario Research Fund, Research Excellence Program Round-9 (ORF-RE09) from the Ontario Ministry of Economic Development, Job Creation and Trade, Canada (Project #053970 and # 054345); and v) the Agriculture and Agri-Food Canada (AAFC), Maple Leaf Food, Canada and Bank of Montreal (BMO), Canada through Bioindustrial Innovation Canada (BIC) Bioproducts AgSci Cluster Program (Project # 054015, 054449 and 800148).

Circular Economy Driven Biocomposites: A Renewal for Innovation and Commercialization in Sustainable Packaging

Conference: Sustainable Packaging Symposium (SPS)

Year: 2020

Proceeding: Sustainable Packaging Symposium 2020

Group: Abstract Submissions

Session: SPS Tank – Innovation Showcase

Time: Monday, June 29, 2020 - 4:00pm-4:25pm

Authors:

Mohanty, A. K., University of Guelph

Since 1950, plastic waste on land and in ocean has accumulated to more than 7.5 billion tons globally. One of the innovative solutions is in adopting a â€œcircular economyâ€ concept, through utilization of wastes to create new industrial products. Recycled and waste material sources are gaining more interest in green packaging sector, as new technologies are being developed. Composite materials are well suited for the utilization of waste sources, including waste plastics and other bio-based wastes. In addition, a range of compostable biocomposites for food packaging uses are possible to engineer by combining biodegradable plastics with industrial waste and coproducts such as coffee chaff, tomato skin, grape pomace, etc. The major scientific challenge is in compatibilizing the bioplastics and biomass residues, while economically maintaining cost-competitiveness of new products. We developed compostable coffee pods, in close collaboration with our industry partners, which is commercially available today. This success was made possible through coffee industry waste valorization in a biocomposite formulation. Overall, biocomposites have been effectively used in valorization of waste plastics and can be one of the enabling strategies towards a circular economy and improved sustainability.

Acknowledgements: i) Natural Sciences and Engineering Research Council (NSERC), Canada Discovery Grants (Project # 401111); ii) the Ontario Ministry of Agriculture, Food and Rural Affairs (OMAFRA) â€“ University of Guelph, the Bioeconomy Industrial Uses Research Program Theme (Project #030255 and 030361); iii) OMAFRA-University of Guelph Gryphonâ€™s Leading to the Accelerated Adoption of Innovative Research (LAAIR) Program (Project # 030416); iv) the Ontario Research Fund, Research Excellence Program; Round-9 (ORF-RE09) from the Ontario Ministry of Economic Development, Job Creation and Trade, Canada (Project # 053970 and 054345); and v) the Agriculture and Agri-Food Canada (AAFC), Maple Leaf Food, Canada and Bank of Montreal (BMO), Canada through Bioindustrial Innovation Canada (BIC) Bioproducts AgSci Cluster Program (Project # 054015, 054449 and 800148).

David Clark is the Vice President of Sustainability at Amcor

Conference: Sustainable Packaging Symposium (SPS)

Year: 2020

Proceeding: Sustainable Packaging Symposium 2020

Group: Abstract Submissions

Session: Panel Discussion - Sustainable Packaging and the SDGs

Time: Tuesday, June 30, 2020 - 2:35pm-3:00pm

Authors:

Clark, D., Amcor

For the panel discussion Iâ€™d like to focus on the work to address plastic waste â€œ in particular the transition to a more circular economy. These will be directly related to SDG12 â€œ responsible consumption and production, SDG11 â€œ sustainable cities and communities, and SDG14 - life below water.

Degradable Bioplastic Wearable/Packaging for Healthy Environment

Conference: Sustainable Packaging Symposium (SPS)

Year: 2020

Proceeding: Sustainable Packaging Symposium 2020

Group: Abstract Submissions

Session: Technical Session - Public Good

Time: Tuesday, June 30, 2020 - 4:20pm-4:45pm

Authors:

Raja, A. A., University of Hafr AlBatin
Stading, M., RISE, Sweden
Lindman, K., Renova AB, Sweden
Wallgren, K., Sweden
Boldizar, A., Chalmers University of Technology, Sweden

The demand of plastic is extremely high due to its use in rigid packaging, flexible packaging, textile industry, automotive industry, transportation, agriculture, horticulture and pharmaceutical industry. The harmful effects of normal plastic waste drove industry to produce non-degradable bioplastics as well as degradable bioplastics; and, as of 2018, bioplastic market value was USD 6.04 billion, reported by Future Business Insightsâ„¢. Over 55% of this bioplastic produced is still non-degradable. The pandemic of corona virus disease, COVID-19, forced people to use disposable plastic (for gloves, bags, ready to eat food, etc). And, disposable plastic clothes became inevitable for medical personnel. The abrupt demand for disposable plastic and packaging to protect oneself from COVID-19 disease, not only increased plastic production but also assisted to generate infected plastic waste. The COVID-19 infected plastic waste urge to envisage research on hydrophobic degradable bioplastic. As non-degradable plastic hinder degradation in composting, the degradable bioplastic should preferably be made from renewable resources. A wide range of bioplastic films were prepared from acetylated high amylose starch with triacetin (AHAST) to be analyzed. The AHAST films had an average thickness of 41.33 microns. AHAST films were subjected to material strength analysis, water absorption capacity analysis and loss of mass analysis at compost facility. The morphological analysis was conducted by SEM to observe physical changes at microscopic level. AHAST films showed 1029 MPa of Young Modulus with 0.11mm/mm elongation at break. With no water uptake, hydrophobic nature of the AHAST bioplastic was evidenced. The AHAST film of 110 mg was composted to smaller fragments of 90 mg in a short period of ten weeks only. The compost facility was comprised of warmth humid closed stage and outdoor aerated piles stage. It was, therefore, concluded that AHAST bioplastic films found to be hydrophobic to serve COVID-19 requirement for disposable plastic and packaging as well as degraded in short period of time to get rid of COVID-19 infected plastic waste.

Design of Bioplastics Based Films By Solvent Casting

Conference: Sustainable Packaging Symposium (SPS)

Year: 2020

Proceeding: Sustainable Packaging Symposium 2020

Group: Abstract Submissions

Session: Technical Session - Novel Products and Technology

Time: Thursday, July 2, 2020 - 2:10pm-2:35pm

Authors:

Gupta, S., University of North Dakota

There is an urgent need for novel materials design solutions which can potentially replace fossil fuel derived plastics. In this presentation, I will report some of the current research which is being done in my research group on bioplastics. More particularly, I will focus on renewable and biodegradable bioplastics derived from plant- and bacteria-based feed stock. As a part of this study, I will review some of the research being done on Poly Lactic Acid (PLA), Poly Hydroxyalkanoate (PHA), Cellulose Acetate (CA), and lignin based compositions. Different types of processing techniques drop-on-demand, Fused Deposition Modeling (FDM), and Solvent Casting will be reviewed. The microstructure, thermal, and mechanical behavior of different types of blends will be reported. In addition, the effect of different particulates like ternary laminates (MAX Phases) and/or phase separation of the blend on thermo-mechanical behavior will be explained. It is hypothesized that these materials can be used for active packaging applications.

Topics:

Developing CarbonSmartTM Packaging Solutions

Conference: Sustainable Packaging Symposium (SPS)

Year: 2020

Proceeding: Sustainable Packaging Symposium 2020

Group: Abstract Submissions

Session: Technical Session - Novel Products and Technology

Time: Thursday, July 2, 2020 - 3:25pm-3:50pm

Authors:

Brenc, R., LanzaTech

We continue to need carbon to make the materials and chemicals needed by society. The manufacture of these goods results in an abundance of waste carbon in the form of polluting gases or solid wastes that end up in the air, our landfill, or the oceans.

The current plastic crisis has led to an increased urgency to establish effective strategies for re/upcycling plastics and other polymers. According to the EPA, the U.S. produces 35.3 million tons of plastics per year, of which only 8% is recycled while the remaining 26.8 million tons end up in landfills. With current technology only a small proportion of this material can be economically sorted and recycled into a product that is functionally equivalent, equal to or better than the input plastic. Utilizing their commercialized gas fermentation platform, LanzaTech is developing multiple novel routes to manufacture CarbonSmart™ packaging from recycled carbon at scale.

The LanzaTech gas fermentation platform recycles carbon rich wastes, including waste emissions from industry and gasified household wastes, into the chemical building blocks needed for packaging production. By reusing waste carbon, packaging that would otherwise come from fresh fossil resources can now be produced from wastes and residues without any loss of properties. Waste carbon that would otherwise fill our air, our oceans, or end up in the landfill, can be recycled into new goods and at the end of its useful life, going through the same carbon recycling process again (and again and again). This presentation will highlight how chemical recycling to produce packaging embodies the circular economy by locking waste carbon in the material cycle.

Extraction of Sugars, Pectic Hydrocolloids, Flavonoids and Peel Oil from Orange Juice Processing Residues By Steam Explosion and Their Potential Value

Conference: Sustainable Packaging Symposium (SPS)

Year: 2020

Proceeding: Sustainable Packaging Symposium 2020

Group: Abstract Submissions

Session: Technical Session - Novel Products and Technology

Time: Thursday, July 2, 2020 - 1:45pm-2:10pm

Authors:

Dorado, C., Agricultural Research Service, USDA
Cameron, R. G., USDA
Manthey, J. A., USDA

In the 2016-2017 season, 96 percent or 2.7 million metric tons of oranges produced in the state of Florida were processed into juice leaving behind 1.2 million metric tons of peel, rag, juice sacs and seeds. These orange juice processing residues are currently pressed and dried into low value cattle feed but contain sugars, pectic hydrocolloids, flavonoids and peel oil that if extracted, can be used in high value products. Therefore, we investigated the use of steam explosion for extraction of sugars, pectic hydrocolloids, flavonoids and peel oil from the two major Florida orange juice processing varieties and calculated their potential value. Early and late harvest samples of Hamlin and Valencia juice processing residues from the 2016-2017 season were subjected to steam explosion in a batch system at 130, 150 and 170 degrees Celsius for 1,2,4, and 8 minute hold times. The resulting steam exploded material was then analyzed for sugars, pectic hydrocolloids, flavonoids and peel oil. In general, more than half of the available glucose and fructose was extracted at the highest temperature and longer hold times. This was also the case for the flavanone glycoside, hesperidin. Steam explosion has the potential to volatilize up to 94 percent of the available peel oil. Galacturonic acid, the major constituent of pectic hydrocolloids, could be recovered at higher temperatures and shorter hold times but with increased molecular weight due to suspected aggregation. It was determined that millions of dollars of potential value exist in the residues based on the amount of glucose, hesperidin, peel oil, and galacturonic acid that could be extracted by steam explosion from the residues produced during the 2016-2017 Florida citrus season.

Extrusion Coatings: Integrating Paper Fibers into Flexible Packaging Designs

Conference: Sustainable Packaging Symposium (SPS)

Year: 2020

Proceeding: Sustainable Packaging Symposium 2020

Group: Abstract Submissions

Session: SPS Tank – Innovation Showcase

Time: Monday, June 29, 2020 - 3:10pm-3:35pm

Authors:

Gasparik, T., Smart Planet Technologies

Globally, millions of pounds of plastic are used in flexible film laminations that are subsequently converted into pouches, sachets, form-fill-seal, bags, sacks, etc. Therefore, flexible packaging has a strategic impact on stakeholders, the environment, and the packaging supply chain. This module showcases success stories e.g. how paper fibers and 40% to 51% mineralized resin(s) provide alternatives to traditional plastic laminations. Also explained is how these novel structures(s) reduce plastic, sustain barrier performance, and expand recycling streams. As explored are design and heat-seal characteristics making it possible to form packaging on high speed converting.

Filling the measurement gap on corporate action on plastic waste

Conference: Sustainable Packaging Symposium (SPS)

Year: 2020

Proceeding: Sustainable Packaging Symposium 2020

Group: Abstract Submissions

Session: Invited Talks - Monitoring, Metrics and Standards

Time: Wednesday, July 1, 2020 - 2:05pm-2:30pm

Authors:

Simon, E., World Wide Fund for Nature

While many companies are stepping up to address the global waste crisis with large-scale plastic commitments, they havenâ€™t been equipped with the industry-wide standard to measure how their actions are amounting to impact. Thatâ€™s why World Wildlife Fund (WWF) launched ReSource: Plastic as a first-of-its-kind effort to fill this gap, creating a common measurement framework to quantify corporate impact on plastic waste reduction. In this talk, WWFâ€™s Erin Simon deep dives into how ReSourceâ€™s innovative methodology is not only framing progress, but through an unprecedented step towards corporate transparency is also fostering much-needed collaboration across the industry.

Functional Oxygen Barriers for Food Packaging from Biomass-Derived Materials

Conference: Sustainable Packaging Symposium (SPS)

Year: 2020

Proceeding: Sustainable Packaging Symposium 2020

Group: Abstract Submissions

Session: Technical Session - Novel Products and Technology

Time: Thursday, July 2, 2020 - 2:35pm-3:00pm

Authors:

Meredith, C., Georgia Institute of Technology
Satam, C. C., Georgia Institute of Technology
Ji, Y., Georgia Institute of Technology
Waters, S. L., Georgia Institute of Technology
Lim, E., Georgia Institute of Technology
Shofner, M. L., Georgia Institute of Technology

There is a strong drive to discover and develop alternatives to conventional plastics that offer the ability to be manufactured and used in a circular manner. In a circular economy, as opposed to a linear one, materials are derived from renewable resources or recycled content, and at the end of life, they are able to be circulated back into production via a chemical, physical or biological pathway. A critical need exists to develop such materials for plastic packaging, which represents the largest contributor to unrecyclable or difficult-to-recycle plastic waste. This seminar will describe recent innovations in preparation of barrier films and coatings, suitable for food or pharmaceutical packaging, based on combinations of cellulose- and chitin-based nanomaterials. Cellulose nanocrystals (CNCs) and chitin nanofibers (ChNFs) are oppositely-charged, water-dispersible fibers that are abundant and could form the basis of a platform of renewable, high-performance materials. Because of their high crystalline content, they are of interest for use in barrier packaging. The combination of CNCs and ChNFs to produce alternating layered materials, direct blends, and other composites, results in high barrier properties to oxygen, driven by electrostatics and hydrogen bonding-driven densification. Current challenges are discussed as well, including the improvement to sensitivity to water vapor and manufacturability.

Impacts of ISO Circular Economy Standard on Sustainable Packaging Metrics

Conference: Sustainable Packaging Symposium (SPS)

Year: 2020

Proceeding: Sustainable Packaging Symposium 2020

Group: Abstract Submissions

Session: Invited Talks - Monitoring, Metrics and Standards

Time: Wednesday, July 1, 2020 - 1:40pm-2:05pm

Authors:

Levy, M., First Environment

Circular economy is a concept that is gaining traction on a global basis. Circular thinking emphasizes redesigning products and packaging in terms of reuse, recycling, and repurposing up front, with the goal of eliminating traditional disposal at the end of life. Many say now is the time to transition from a linear economy to an economy based on circularity.

In May 2019, the process of developing an ISO international standard for circular economy began with an inaugural meeting led by the French standardization association (AFNOR). Since then, 67 countries including the U.S. are now participating members of ISOâ€™s Technical Committee (TC) 323, which is developing this global circular economy standard. The U.S. delegation was formed under the American National Standards Institute (ANSI) and the American Society for Testing and Materials (ASTM) in 2019, consists of over 30 members, and is led by elected co-chairs Alison Conroy of Georgia Pacific and Mike Levy of First Environment. The U.S. delegation has designated experts in all four Working Groups established to develop this standard.

This presentation will highlight some of the key discussions and concepts that may impact sustainable packaging and which are being reviewed in the four Working Groups established under TC 323. In particular, Working Group 3, Measuring Circularity Framework, has an approved scope and deliverable to:

- Specify a generic framework for measuring and assessing circularity, taking into consideration the sustainability (environmental, social, and economic) impacts of circular activities
- Provide guidance to organizations, sectors, and stakeholder groups regarding how to use the generic framework for sector-specific measurement or stakeholder-specific measurement
- Give initial guidance how specific sectors or stakeholders can use this framework and customize it for their purposes

In addition, this presentation will highlight some of the current discussions around metrics and circular economy and the potential impacts for sustainable packaging in the future. These may involve packaging design changes, process changes, and restructuring recovery efforts to avoid waste streams, all with the goal of â€œclosing the loop.â€ This information will be particularly beneficial to stakeholders considering enacting projects that will comply with the final circular economy ISO standard.

Innovation Pathways for More Circular Plastic Packaging

Conference: Sustainable Packaging Symposium (SPS)

Year: 2020

Proceeding: Sustainable Packaging Symposium 2020

Group: Abstract Submissions

Session: Technical Session - Public Good

Time: Tuesday, June 30, 2020 - 3:30pm-3:55pm

Authors:

Dooley, K., Arizona State U
Hafsa, F., Arizona State U
Basile, G., The Global Kaiteki Center
Buch, R., Arizona State U

Plastic packaging has become ubiquitous and has provided many societal benefits, such as increased food shelf-life. Plastic packaging can have a significant environmental footprint though, including risks from microplastics. Increasing the fraction of plastic packaging that is reused or recycled could reduce these impacts, but only about 10% of plastic packaging is reused or recycled worldwide. Even in developed economies like the US or Europe, more plastic goes to landfill than to recycling.

Our research, sponsored by the Global Kaiteki Center at Arizona State University, has identified over 300 innovations in the public domain that have the potential to make the plastic packaging value chain more circular, i.e. reduce, reuse, or recycle plastic packaging. Over 140 participants from across the plastic value chain, including corporations, government, NGOs, and academics, have reviewed and provided feedback on our list of innovations. We have created a decision-model that helps stakeholders understand how specific bundles of these innovations can increase the percentage of plastic packaging that is reused and recycled.

For example, in order for plastic packaging to end up being recycled, it has to have materials and a design form that can be recyclable â€“ innovations in polymer or biopolymer design help increase the available solution space for recyclable packaging. Further downstream, a consumer must properly dispose of the packaging in a recycling bin for it to eventually be recycled. Innovations aimed at consumer education or communication, like recycling labels, increase awareness and the likelihood the consumer will act correctly. Conversely, innovations in MRF sorting could eliminate the need for the consumer to separate recyclables at all. The presentation will share our model and summarize the insights we can derive from it.

Integration of Waste into Circular Carbon Economy: Better Carbon Management for Climate Change Mitigation and Sustainable Development in the Case of Food Waste

Conference: Sustainable Packaging Symposium (SPS)

Year: 2020

Proceeding: Sustainable Packaging Symposium 2020

Group: Abstract Submissions

Session: Technical Session - Public Good

Time: Tuesday, June 30, 2020 - 3:05pm-3:30pm

Authors:

Guran, S., Rutgers EcoComplex
Deshmukh, R., NJDEP

Integration of Waste into Circular Carbon Economy: Better Carbon Management for Climate Change Mitigation and Sustainable Development in the Case of Food Waste

Serpil Guran*, Rupa Deshmukh**

***Rutgers EcoComplex**

****New Jersey Department of Environmental Protection**

Abstract:

The emerging concept of â€œcircular economyâ€ has attracted interest of various stakeholders with varying definitions and achievement goals. Circular economy can promote sustainable development and positively contribute to carbon footprint reduction and help mitigate climate change. Circular economy can be an effective pathway for achieving lower-carbon economy. Therefore, promoting combined understanding of circularity and lower-carbon economy as â€œ**circular carbon economy**â€ and emphasizing the transformation of linear make-it /use-it/dispose-it pathway to circular resource recovery pathway can provide better results. To date, greenhouse gas (GHG) emissions inventories generally report that emissions related to waste management matters globally may not be as high as other GHG emitting sources i.e. electricity generation from fossil fuel combustion or transportation and may create understanding that innovative approaches in waste management sector may not contribute to climate change mitigation efforts substantially. This paper examines how innovative waste management approaches can serve as an integral component in achieving â€œcircular carbon economyâ€ in New Jersey by 2050.

Key Words: Circular economy, low-carbon economy, waste disposal, resource recovery, climate change mitigation, food waste

Jessica Bowman is an Executive Director at Plant Based Products Council's (PBPC)

Conference: Sustainable Packaging Symposium (SPS)

Year: 2020

Proceeding: Sustainable Packaging Symposium 2020

Group: Abstract Submissions

Session: Panel - Novel Products and Technologies

Time: Thursday, July 2, 2020 - 3:20pm-3:45pm

Authors:

Bowman, J., Plant Based Products Council

As research into more innovative packaging materials continues, it is important to consider the pragmatic barriers these materials may face as they look to enter the market, along with the role innovation can play in tackling some of the technical challenges more sustainable packaging materials encounter.

Keefe Harrison is the Founder and CEO of The Recycling Partnership

Conference: Sustainable Packaging Symposium (SPS)

Year: 2020

Proceeding: Sustainable Packaging Symposium 2020

Group: Abstract Submissions

Session: Panel Discussion - Circularity, reuse, and recycling

Time: Monday, June 29, 2020 - 1:50pm-2:15pm

Authors:

Harrison, K., The Recycling Partnership

Keefe Harrison, CEO of The Recycling Partnership, will discuss the current status of recycling in the US and what will be needed to deliver a stronger system. The Recycling Partnership is rooted in data-based solutions to advance the circular economy of tomorrow.

Making Sense of the Data: New Measures for a Circular Economy

Conference: Sustainable Packaging Symposium (SPS)

Year: 2020

Proceeding: Sustainable Packaging Symposium 2020

Group: Abstract Submissions

Session: Plenary Talk 7/1

Time: Wednesday, July 1, 2020 - 1:05pm-1:30pm

Authors:

Martin, E., The Circulate Initiative

As the public and private sectors seek solutions to stop the flow of plastic waste into the ocean, the need for better measures is clear. However, itâ€™s a â€œWild Westâ€ moment for measurement. There are actually many resources out there on what and how to measure, but itâ€™s fragmented and not accessible. Assurances for tracking and tracing materials through a post-consumption supply chain are emerging, but not yet at scale. If we want to be able to act quickly, we need to do better. This session will provide an overview of the measurement landscape and what's needed in the industry.

Many Happy Returns - How Refillable and Returnable Packaging Systems Could Solve the Plastics Problem

Conference: Sustainable Packaging Symposium (SPS)

Year: 2020

Proceeding: Sustainable Packaging Symposium 2020

Group: Abstract Submissions

Session: Technical Session - Public Good

Time: Tuesday, June 30, 2020 - 3:55pm-4:20pm

Authors:

Greenwood, S., University of Sheffield
Rothman, R. H., University of Sheffield
Parsons, R., University of Sheffield
Baird, H., University of Sheffield
Walker, S., University of Sheffield
Ryan, A., University of Sheffield
Jackson, P., University of Sheffield
Webb, T., University of Sheffield
Slark, A., University of Sheffield

An estimated 2.26m tonnes single-use plastic packaging goes to waste each year in the UK. The introduction of circular reusable packaging systems has the potential to significantly reduce plastics waste and reduce the overall environmental impact of the packaged product.

That packaging should be recyclable, recoverable and/or reusable has been part of EU and UK regulations since 1996. To date, reuse has largely been limited to secondary packaging (e.g.

crates), some primary packaging (e.g. bottles for beverages) and niche systems (e.g. refills in zero-waste shops), although the food retail sector is increasingly implementing reuse initiatives. The New Plastics Economy Global Commitment and the UK Plastics Pact launched with the aim of making plastic packaging more circular, including specific targets for Reusable packaging systems.

Barriers to and facilitators of reuse, from both technical and behavioral perspectives, have been identified. Products where consumers are most likely to want to reuse the packaging have been determined and reuse behavior and attitudes to reuse studied. We focus on the traditional example of doorstep milk delivery, reflecting on the potential barriers and enablers to making reuse mainstream. The case highlights that the transition from reuse to single-use, and therefore a potential reverse, is entangled in a complex network of technical, social and economic factors. The life cycle analysis of a range of durable plastic materials have been investigated to find suitability for reuse and ensure that proposed reuse systems are truly sustainable over their entire life cycle, and that any additional emissions do not outweigh the benefits.

PHA: Developing Practical Biodegradable Single-Use Plastics

Conference: Sustainable Packaging Symposium (SPS)

Year: 2020

Proceeding: Sustainable Packaging Symposium 2020

Group: Abstract Submissions

Session: SPS Tank – Innovation Showcase

Time: Monday, June 29, 2020 - 3:35pm-4:00pm

Authors:

Mang, M., Danimer Scientific

Consumers want to live more sustainable lives including reducing the impact of plastic items that they use on the environment. Danimerâ€™s Nodax PHA materials can be used to make practical single use plastic items with end-of-life characteristics that can be tailored to fit designed or accidental disposal systems. The product technology highlighted today is profile extrusion to produce marine biodegradable drinking straws. The rapid development of a material that performs and has been commercialized will be described

Plant-based Packaging - Understanding the Opportunities and Challenges

Conference: Sustainable Packaging Symposium (SPS)

Year: 2020

Proceeding: Sustainable Packaging Symposium 2020

Group: Abstract Submissions

Session: Plenary Talk 7/2

Time: Thursday, July 2, 2020 - 1:05pm-1:30pm

Authors:

Bowman, J., Plant Based Products Council

As consumers and brands look for more sustainable packaging options, interest in the use of plant-based materials continues to grow. Using packaging materials made from renewable resources, rather than traditional petroleum-based materials, can lead to numerous environmental benefits, from reduced greenhouse gas emissions to improvements in soil health. These often nascent products are not without their challenges though -- lack of awareness, questions about performance, cost, and availability of end-of-life infrastructure to name a few. Tune in to this plenary talk to hear about the potential benefits of plant-based packaging, along with challenges to be tackled in order to realize the full circular potential of these materials.

Plastics, Plastic Packaging, and the SDGs

Conference: Sustainable Packaging Symposium (SPS)

Year: 2020

Proceeding: Sustainable Packaging Symposium 2020

Group: Abstract Submissions

Session: Plenary Talk 6/30

Time: Tuesday, June 30, 2020 - 1:05pm-1:30pm

Authors:

Clark, D., Amcor

Plastics and plastic packaging have come under pressure because of concern about plastic waste. Many consumers and government officials are rightly concerned about the negative impacts that plastic waste has on our environment. Many people donâ€™t realize the benefits that plastic packaging offer to society and to the environment, which are well summarized in terms of the Sustainable Development Goals.

Packaging plays an important role in delivering food, medicine, cleaning suppliers, and other products to the people who need them most, contributing directly to reducing hunger and promoting good health while also reducing waste. Plastics are often more efficient than alternative materials, reducing GHG emissions and other impacts. Work being done by many companies as part of the Global Commitment to End Plastic Waste is moving the industry toward more circular-economy solutions. These will help move our industry and society toward more responsible consumption and production, promote more sustainable cities and communities, and improve life below water.

Prapti Muhuri is the Manager of Recycling and Recovery at the American Chemistry Council (ACC)

Conference: Sustainable Packaging Symposium (SPS)

Year: 2020

Proceeding: Sustainable Packaging Symposium 2020

Group: Abstract Submissions

Session: Panel Discussion - Circularity, reuse, and recycling

Time: Monday, June 29, 2020 - 2:15pm-2:40pm

Authors:

Muhuri, P., American Chemistry Council

Prapti Muhuri of the American Chemistry Council will discuss the opportunities for increasing circularity, recent growth in advanced recycling and the role that mass balance certification could play in accelerating plastics recycling and increasing recycled content in packaging.

Recycling and Packaging: The Intersection of Two Industries

Conference: Sustainable Packaging Symposium (SPS)

Year: 2020

Proceeding: Sustainable Packaging Symposium 2020

Group: Abstract Submissions

Session: Plenary Talk 6/29

Time: Monday, June 29, 2020 - 1:05pm-1:30pm

Authors:

Robinson, S., Waste Management

From the beginning of time, humans have recycled -initially out necessity, but eventually, recycling became a global trade supplying feedstock to the manufacturing industry. By the 20th century, markets for metals and paper ebbed and flowed in a commodity-based market system.

While commercial and industrial recycling still functions in much the same way, in the 1990â €™s local policies began to emphasis the growth of residential recycling programs as a solid waste disposal alternative. Residential recycling programs focused on building supply without corresponding policies to create sufficient demand. As a result, recyclables moved to countries that needed the feedstock for their growing economics. For several decades, as residential recycling programs grew, export demand kept our supply and demand in balance.

Fast forward to 2017. Chinaâ€™s changing policies to clean up their environment disrupted global recycling markets. The result was a reduction of 50% of the worldâ€™s demand for mixed paper and mixed plastic, which drove down global commodity prices. Ironically, this led to new technology investments to improve the quality of recyclable feedstock and to improve efficiency in order to improve the economics of recycling.

As the pandemic hit, MRFs that had made these investments saw their investments pay off with systems that could continue to function in a COVID environment.

In this session, weâ€™ll talk about the role of packaging design, recycling technology, material quality, policy and economics -drawing from past experiences to imagine our vision for the future.

Revolutionizing Materials. Right. Now

Conference: Sustainable Packaging Symposium (SPS)

Year: 2020

Proceeding: Sustainable Packaging Symposium 2020

Group: Abstract Submissions

Session: Technical Session - Circularity, reuse, and recycling

Time: Wednesday, July 1, 2020 - 2:35pm-3:00pm

Authors:

Leita, J., Eastman

Many major brands now have established goals for recycled, sustainable solutions for their products and packaging. But how will they make the transition? Traditional mechanical recycling is inefficient. Most chemical recycling solutions are still several years out. Until now. Eastman is a leader in todayâ€™s circular economy. We have reimagined our material production process and developed a way to replace the coal and natural gas used for feedstocks with plastic waste currently destined for the landfill. Our presentation will cover how we create materials made from waste plastic that can be used to craft packaging and products right now â€” not in the distant future.

Steve Alexander is the President and CEO of the Association of Plastic Recyclers (APR)

Conference: Sustainable Packaging Symposium (SPS)

Year: 2020

Proceeding: Sustainable Packaging Symposium 2020

Group: Abstract Submissions

Session: Panel Discussion - Circularity, reuse, and recycling

Time: Monday, June 29, 2020 - 2:40pm-3:05pm

Authors:

Alexander, S., Association of Plastic Recyclers (APR)

Steve Alexander will address the imperative and market implications of Design for Recyclability; PCR Certification; and Demand as the key components to sustainability for plastics packaging.

Sustainable Packaging for the Circular Food Economy

Conference: Sustainable Packaging Symposium (SPS)

Year: 2020

Proceeding: Sustainable Packaging Symposium 2020

Group: Abstract Submissions

Session: Panel - Novel Products and Technologies

Time: Thursday, July 2, 2020 - 3:45pm-4:10pm

Authors:

Trabold, T., RIT
Diaz-Acosta, C., Rochester Institute of Technology
Lewis, C., RIT
Hegde, S., RIT

The impacts of food waste on the U.S. economy are staggering, accounting for 21% of fresh water use, 19% of fertilizer use, 18% of cropland, and over 21% of landfill volume, all at an annual cost in excess of $200 billion. Despite the current focus on food waste and methods for source reduction and landfill diversion, very little is known about the associated problem of food packaging waste (FPW). Although FPW is a subset of the much broader food waste issue, it represents an important topic requiring immediate focus, for two reasons. First, there is very little known about how much FPW is generated across the food supply chain, and published studies on this topic are almost nonexistent. To develop a broad food waste management strategy that will result in landfill diversion on a large scale, FPW needs to also be addressed so that value-added alternatives can be developed. Secondly, FPW represents a significant technical challenge because it is part of a â€œsystemâ€ comprised of a biodegradable, organic fraction (i.e., the food material itself) combined with packaging material that may be recyclable, but in many cases is non-biodegradable (e.g., plastic) and thus different valorization methods are needed. It is desirable to develop entirely new methods for food packaging that essentially eliminates waste by using these materials at the end of life as value-added inputs to other stages of the food supply chain, thus enabling circularity and avoiding the takeâ†'makeâ†'waste practice of the conventional linear economy.

In this research, we explore two approaches for developing food packaging that supports a circular food economy model. The first involves bioplastic packaging materials designed to have accelerated rates of anaerobic digestion exceeding those of commercially available materials, using methods including co-polymerization, blending, and mineral fillers. Preliminary experiments also suggest there can be strong interactions between certain bioplastic materials and food waste, and we hypothesize that the food waste type (e.g., pre-consumer, post-consumer, etc.) may also play a role in accelerating the anaerobic degradation of these new bioplastic materials. The second approach involves development of composite packaging containing biochar, carbonaceous material produced by the thermochemical conversion of organic matter, including food waste. Thermoformed containers were manufactured with 10, 20 and 30 wt.% biochar derived from waste coffee grounds in a mixture of thermoplastic starch (TPS) and polycaprolactone (PCL). The addition of biochar decreased the elongation at break, but did not significantly affect the modulus of elasticity or tensile strength. Preliminary experiments also indicate that adding 10 wt.% biochar increased the biodegradation of the material. These approaches have motivated other innovations that may improve the sustainability of plastic use across the food supply chain, such as development of new biochar-bioplastic composite mulching films that fully degrade at the end of the growing season, and are designed to return organic carbon to the soil.

Transforming systems of consumption and production for achievingthe sustainable development goals: moving beyond efficiency

Conference: Sustainable Packaging Symposium (SPS)

Year: 2020

Proceeding: Sustainable Packaging Symposium 2020

Group: Abstract Submissions

Session: Panel Discussion - Sustainable Packaging and the SDGs

Time: Tuesday, June 30, 2020 - 3:00pm-3:25pm

Authors:

Cohen, M., Search Results Web Result with Site Links New Jersey Institute of Technology

The United Nations formulated the sustainable development goals (SDGs) in 2015 as a comprehensive global policy framework for addressing the most pressing social and environmental challenges currently facing humanity. In this paper, we analyse SDG 12, which aims to â€˜â€˜ensure sustainable consumption and production patterns.â€™â€™ Despite long-standing political recognition of this objective, and ample scientific evidence both on its importance and on the efficacy of various ways of promoting it, the SDGs do
not provide clear goals or effective guidance on how to accomplish this urgently needed transformation. Drawing from the growing body of research on sustainable consumption and production (SCP), the paper identifies two dominant vantage pointsâ€” one focused on promoting more efficient production methods and products (mainly through technological improvement and informed consumer choice) and the other stressing the need to consider also overall volumes of consumption, distributional issues, and related social and institutional changes. We label these two approaches efficiency and systemic. Research shows that while the efficiency approach contains essential elements of a transition to sustainability, it is by itself

highly unlikely to bring about sustainable outcomes. Concomitantly, research also finds that volumes of consumption and production are closely associated with environmental impacts, indicating a need to curtail these volumes in ways that safeguard social sustainability, which is unlikely to be possible without a restructuring of existing socioeconomic arrangements. Analysing how these two perspectives are reflected in the SDGs framework, we find that in its current conception, it mainly relies on the efficiency approach. On the basis of this assessment, we conclude that the SDGs represent a partial and inadequate conceptualisation of SCP which will hamper implementation. Based on this determination, this paper provides some suggestions on how governments and other actors involved in SDGs operationalisation could more effectively pursue SCP from a systemic standpoint and use the transformation of systems of consumption and production as a lever for achieving multiple sustainability objectives.

Valorization of Waste Fines and Reject Fibers Produced in Recycled Liner Manufacturing - Biochemical Conversion Processes

Conference: Sustainable Packaging Symposium (SPS)

Year: 2020

Proceeding: Sustainable Packaging Symposium 2020

Group: Abstract Submissions

Session: Technical Session - Circularity, reuse, and recycling

Time: Wednesday, July 1, 2020 - 3:50pm-4:15pm

Authors:

Ramarao, B. V., ESPRI, SUNY College of Environmental Science and Forestry

Recycled linerboard mills produce packaging paper from old corrugated carton (OCC) wastes by first repulping them, followed by refining of the pulps to improve the bonding abilities of the fibers. A large amount of fiber fragments known as fines are produced in this process. The waste streams contain significant amount of such fines in addition to calcium carbonate and clay fillers. The removal of hornified cellulosic fines can accelerate water removal from the pulps on the papermachines and in the dryer sections, leading to significant productivity increases while improved inter-fiber bonding in the sheet due to a lower â€˜waste finesâ€™ content improves strength. These waste fines and fiber rejects from the repulping process can be converted into higher value added products if successful conversion processes are developed.

We discuss the bioconversion of these waste fines by enzymatic hydrolysis into sugars which are further converted into biodegradable bioplastics such as polyhydroxy-alkanoates (PHA)s. Other routes for valorization can include methods of 're-activating' the hornified cellulosic surfaces by cationic polymers and their reintroduction into the furnish.

Biodegradable Mulch Film from Soy Waste

Conference: Sustainable Packaging Symposium (SPS)

Year: 2020

Proceeding: Sustainable Packaging Symposium 2020

Group: Poster Submissions

Session: Poster Session

Authors:

Farfaras, N., University of Massachusetts Lowell

We continue to need carbon to make the materials and chemicals needed by society. The manufacture of these goods results in an abundance of waste carbon in the form of polluting gases or solid wastes that end up in the air, our landfill, or the oceans.

The current plastic crisis has led to an increased urgency to establish effective strategies for re/upcycling plastics and other polymers. According to the EPA, the U.S. produces 35.3 million tons of plastics per year, of which only 8% is recycled while the remaining 26.8 million tons end up in landfills. With current technology only a small proportion of this material can be economically sorted and recycled into a product that is functionally equivalent, equal to or better than the input plastic. Utilizing their commercialized gas fermentation platform, LanzaTech is developing multiple novel routes to manufacture CarbonSmart™ packaging from recycled carbon at scale.

The LanzaTech gas fermentation platform recycles carbon rich wastes, including waste emissions from industry and gasified household wastes, into the chemical building blocks needed for packaging production. By reusing waste carbon, packaging that would otherwise come from fresh fossil resources can now be produced from wastes and residues without any loss of properties. Waste carbon that would otherwise fill our air, our oceans, or end up in the landfill, can be recycled into new goods and at the end of its useful life, going through the same carbon recycling process again (and again and again). This presentation will highlight how chemical recycling to produce packaging embodies the circular economy by locking waste carbon in the material cycle.

Effect of Isosorbide Concentration on Alkaline Hydrolysis of Co-Polyesters, Peit

Conference: Sustainable Packaging Symposium (SPS)

Year: 2020

Proceeding: Sustainable Packaging Symposium 2020

Group: Poster Submissions

Session: Poster Session

Authors:

Das, A., University of Toledo

Circular economy is a concept that is gaining traction on a global basis. Circular thinking emphasizes redesigning products and packaging in terms of reuse, recycling, and repurposing up front, with the goal of eliminating traditional disposal at the end of life. Many say now is the time to transition from a linear economy to an economy based on circularity.

In May 2019, the process of developing an ISO international standard for circular economy began with an inaugural meeting led by the French standardization association (AFNOR). Since then, 67 countries including the U.S. are now participating members of ISOâ€™s Technical Committee (TC) 323, which is developing this global circular economy standard. The U.S. delegation was formed under the American National Standards Institute (ANSI) and the American Society for Testing and Materials (ASTM) in 2019, consists of over 30 members, and is led by elected co-chairs Alison Conroy of Georgia Pacific and Mike Levy of First Environment. The U.S. delegation has designated experts in all four Working Groups established to develop this standard.

This presentation will highlight some of the key discussions and concepts that may impact sustainable packaging and which are being reviewed in the four Working Groups established under TC 323. In particular, Working Group 3, Measuring Circularity Framework, has an approved scope and deliverable to:

- Specify a generic framework for measuring and assessing circularity, taking into consideration the sustainability (environmental, social, and economic) impacts of circular activities
- Provide guidance to organizations, sectors, and stakeholder groups regarding how to use the generic framework for sector-specific measurement or stakeholder-specific measurement
- Give initial guidance how specific sectors or stakeholders can use this framework and customize it for their purposes

In addition, this presentation will highlight some of the current discussions around metrics and circular economy and the potential impacts for sustainable packaging in the future. These may involve packaging design changes, process changes, and restructuring recovery efforts to avoid waste streams, all with the goal of â€œclosing the loop.â€ This information will be particularly beneficial to stakeholders considering enacting projects that will comply with the final circular economy ISO standard.

Multiblock Copolymers for Compatibilizing and Recycling PET and PE

Conference: Sustainable Packaging Symposium (SPS)

Year: 2020

Proceeding: Sustainable Packaging Symposium 2020

Group: Poster Submissions

Session: Poster Session

Authors:

Peng, X., University of Minnesota
Ellison, C. J., University of Minnesota
Nomura, K., University of Minnesota
Jin, K., University of Minnesota
Kim, H., University of Minnesota
Kim, H.
Bratton, A. F., Murray State University
Bond, C. R., University of Minnesota
Broman, A. E., University of Minnesota
Miller, K., Murray State University

We continue to need carbon to make the materials and chemicals needed by society. The manufacture of these goods results in an abundance of waste carbon in the form of polluting gases or solid wastes that end up in the air, our landfill, or the oceans.

The current plastic crisis has led to an increased urgency to establish effective strategies for re/upcycling plastics and other polymers. According to the EPA, the U.S. produces 35.3 million tons of plastics per year, of which only 8% is recycled while the remaining 26.8 million tons end up in landfills. With current technology only a small proportion of this material can be economically sorted and recycled into a product that is functionally equivalent, equal to or better than the input plastic. Utilizing their commercialized gas fermentation platform, LanzaTech is developing multiple novel routes to manufacture CarbonSmart™ packaging from recycled carbon at scale.

The LanzaTech gas fermentation platform recycles carbon rich wastes, including waste emissions from industry and gasified household wastes, into the chemical building blocks needed for packaging production. By reusing waste carbon, packaging that would otherwise come from fresh fossil resources can now be produced from wastes and residues without any loss of properties. Waste carbon that would otherwise fill our air, our oceans, or end up in the landfill, can be recycled into new goods and at the end of its useful life, going through the same carbon recycling process again (and again and again). This presentation will highlight how chemical recycling to produce packaging embodies the circular economy by locking waste carbon in the material cycle.

Recent Advances of Corn-Based Polymeric Materials for Packaging: Life Cycle Thinking

Conference: Sustainable Packaging Symposium (SPS)

Year: 2020

Proceeding: Sustainable Packaging Symposium 2020

Group: Poster Submissions

Session: Poster Session

Authors:

Wu, N., North Dakota State University
Pourhashem, G., North Dakota State University

Plastic packaging has become ubiquitous and has provided many societal benefits, such as increased food shelf-life. Plastic packaging can have a significant environmental footprint though, including risks from microplastics. Increasing the fraction of plastic packaging that is reused or recycled could reduce these impacts, but only about 10% of plastic packaging is reused or recycled worldwide. Even in developed economies like the US or Europe, more plastic goes to landfill than to recycling.

Our research, sponsored by the Global Kaiteki Center at Arizona State University, has identified over 300 innovations in the public domain that have the potential to make the plastic packaging value chain more circular, i.e. reduce, reuse, or recycle plastic packaging. Over 140 participants from across the plastic value chain, including corporations, government, NGOs, and academics, have reviewed and provided feedback on our list of innovations. We have created a decision-model that helps stakeholders understand how specific bundles of these innovations can increase the percentage of plastic packaging that is reused and recycled.

For example, in order for plastic packaging to end up being recycled, it has to have materials and a design form that can be recyclable â€“ innovations in polymer or biopolymer design help increase the available solution space for recyclable packaging. Further downstream, a consumer must properly dispose of the packaging in a recycling bin for it to eventually be recycled. Innovations aimed at consumer education or communication, like recycling labels, increase awareness and the likelihood the consumer will act correctly. Conversely, innovations in MRF sorting could eliminate the need for the consumer to separate recyclables at all. The presentation will share our model and summarize the insights we can derive from it.

AUTHOR INDEX